MYSTERIES & MARVELS OF INSECT LIFE

Dr. Jennifer Owen

Edited by Rick Morris

Designed by Anne Sharples
and Teresa Foster

Illustrated by Ian Jackson
and Alan Harris

Cartoons by John Shackell

First published in 1984
by Usborne Publishing Ltd, 20 Garrick Street,
London WC2E 9BJ.
© 1984 by Usborne Publishing Ltd.

Printed in Great Britain

Some shieldbugs, unlike most insect parents, stand guard over their eggs and young.

The 13 cm Goliath Beetle is the heaviest flying insect.

Not thorns, but tree-hoppers.

Contents

A Seven Spot Ladybird to the same scale.

Largest and smallest butterflies drawn life-size: a female Queen Alexandra's Birdwing and a Dwarf Blue.

Assassin bugs inject saliva into their prey, then suck its juices.

Painted Ladies fly over 5,000 km from Africa to Britain.

A 12 mm click beetle can catapult to a height of 300 mm.

About this book

There are about a million species of insects and they outnumber all other sorts of animals by almost four to one. This book is a stimulating introduction to their extraordinary variety and abundance, and concentrates on the more curious and unexpected aspects of insect life.

Insects range in size from larger than the smallest mammals to those tiny enough to crawl through the eye of a needle. The following pages describe some of the ingenious structures and strategies that they use in staying alive — moving, feeding and defending themselves — and in reproducing their kind. The book looks at beetles that stick like limpets, blood-sucking moths, praying mantids that look like flowers, sawflies that vomit pine resin at attackers, the language of bees, and the link between poisons and colours.

An insect's body is in three parts and all insects have three pairs of legs. Some do not have a common name and are called by their latin-based scientific name.

This book will lead to an understanding of why insects are so successful but also points out that there is still much to be discovered about insects and the way they live.

Well-camouflaged Peppered Moths: the dark form has developed where pollution blackens tree trunks.

An Ant-lion larva digs a trap for ants.

Ground level

A cut-away view of the trap

A leaf insect, complete with leaf-like marks and legs like twigs.

TRUE or FALSE?

Look out for these questions and try to guess if they are true or false. The answers are on p. 32.

Ingenious design

The bodies of insects are amazingly varied in shape and form. All these designs are answers to the problems of moving, breathing, keeping warm, eating and avoiding being eaten.

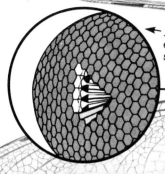

← A close-up of the eye showing the six-sided lenses.

Between the compound eyes are three small simple eyes which respond to light.

Aerial hunters ▶

The large second and third segments of a dragonfly's body are tilted, bringing all the legs forward below the jaws, where they form a basket for catching other insects in flight. Dragonflies strike with deadly accuracy, guided by the enormous compound eyes, each made up of as many as 30,000 separate lenses. Each lens reflects a slightly different view of the world.

Dragonfly hunting a butterfly.

Water-skiing

In an emergency, the rove beetle, *Stenus*, only 5 mm long, can zoom across the surface of water. Glands at the tip of the abdomen release a liquid that lowers the surface tension of water. The beetle is pulled forward by the greater surface tension of the water in front of it.

Each foot has 5,000 bristles which end in pairs of flat pads.

▼ A gripping story

When threatened, a tiny leaf-eating beetle, *Hemisphaerota cyanea*, can clamp down like a limpet. Just as water between two sheets of glass sticks them together, so the beetle uses a film of oil between the 60,000 pads on its feet and the leaf.

Ants heave unsuccessfully at the rounded "shell".

Hemisphaerota cyanea

There are over 40,000 species of weevil. They lay eggs in nuts, such as chestnuts, hickory and hazel, which the grub eats from the inside.

Antennae

Jaws

Elephant Weevil

All the colours of the rainbow

Butterfly wings are covered with tiny over-lapping scales. Iridescent colours result from the way some scales reflect light, and depend on structure not pigment. Other scales have colour pigments.

In this close-up of scales, the orange and deep purple ones have pigment colours. The iridescent blue and green scales have little pigment but reflect blue or green light.

Morpho butterfly

Scales increase "lift" when flying.

Two views of an iridescent wing. Colour depends on the light angle.

A boring story ▶

The jaws of weevils are at the end of a long snout which, in nut weevils, may be as long as the rest of the body. The tiny weevils bore holes in hard nutshells by using the snout as a lever to increase pressure.

Scuba diving and air lines

Some diving beetles use an air bubble trapped between their bodies and their wing cases for breathing under water. Oxygen from the water replaces some of the air used but slowly the bubble shrinks and the beetle must re-surface. Other insects have breathing tubes: the Rat-tailed Maggot's tube is telescopic.

The Water Scorpion has an air tube.

Some diving beetles tow an air bubble.

The Rat-tailed Maggot's tube is 27 cm.

TRUE or FALSE?

Bumblebees have central heating.

5

Colourful confusion

Many insects fool predators by looking like something else or by camouflage colours which match their backgrounds.

Bay-headed Tanager

Which way? ▶

Hairstreak butterflies have a dummy head on their hind wings. On landing, they move their wings up and down. This waggles the false antennae and makes them look real. A bird will probably peck at the false head rather than the real one. The butterfly then escapes with only slightly damaged wings.

Head

This South American hairstreak will fly "the wrong way" to escape, confusing the bird.

False antennae

The Mocker Swallowtail ▼

Birds quickly learn the colour patterns of poisonous butterflies and leave them alone. Edible butterflies which have the same patterns are protected. Female Mocker Swallowtails have the added advantage of looking like several poisonous species. Males, strangely, never mimic and so are unprotected.

One model: a poisonous Friar butterfly.

Another model: a poisonous African Monarch.

Two mimics: both are female Mocker Swallowtails.

Curious caterpillars ▶

Caterpillars make juicy meals but protect themselves from birds by various disguises. Some look like bird droppings, others like twigs, and some match the colours of leaves. The Lobster Moth caterpillar even squirts acid.

This threatening "snake" is actually a hawk moth caterpillar. It waves its body to look like a snake.

Not a twig but a caterpillar holding itself rigid.

The false eyes are on the underside.

6

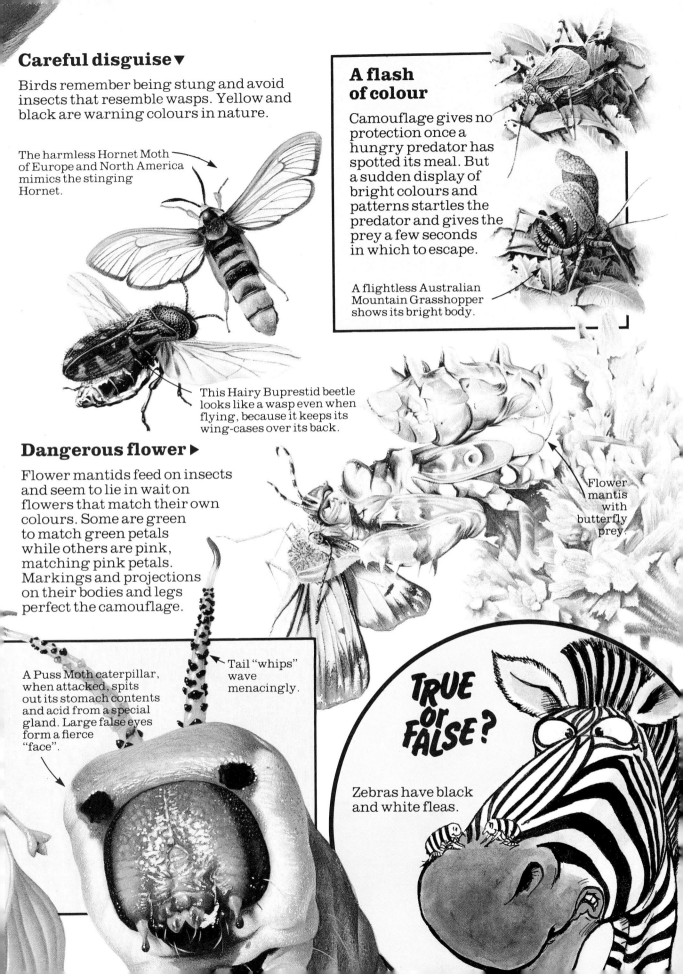

Careful disguise ▼

Birds remember being stung and avoid insects that resemble wasps. Yellow and black are warning colours in nature.

The harmless Hornet Moth of Europe and North America mimics the stinging Hornet.

This Hairy Buprestid beetle looks like a wasp even when flying, because it keeps its wing-cases over its back.

A flash of colour

Camouflage gives no protection once a hungry predator has spotted its meal. But a sudden display of bright colours and patterns startles the predator and gives the prey a few seconds in which to escape.

A flightless Australian Mountain Grasshopper shows its bright body.

Flower mantis with butterfly prey.

Dangerous flower ▶

Flower mantids feed on insects and seem to lie in wait on flowers that match their own colours. Some are green to match green petals while others are pink, matching pink petals. Markings and projections on their bodies and legs perfect the camouflage.

Tail "whips" wave menacingly.

A Puss Moth caterpillar, when attacked, spits out its stomach contents and acid from a special gland. Large false eyes form a fierce "face".

TRUE or FALSE?

Zebras have black and white fleas.

Food and feeding

Insects eat plants or animals, alive or dead, as well as substances produced by them. Some have amazing methods of using unlikely sources of food.

Toad-in-the-hole ▶

The larvae of a horsefly lie buried in mud at the edge of ponds in the U.S.A., ready to grab tiny Spade-foot Toads as they emerge from the water. A larva grasps a toadlet with its mouth hooks and injects a slow-acting poison. Then it drags its meal into the mud and sucks out the juices. It leaves the rest of the body to rot. Adult horseflies, however, are likely to be eaten by adult toads.

Young Spade-foot Toad.

The toadlet and the horsefly larva are each about two centimetres long.

The larva is a grub-like stage between the egg and adult horsefly.

Disappearing dung

Dung beetles bury the dung of grazing mammals and lay their eggs on it. They react to smell, moving towards buffalo dung before it hits the ground. A single mass of fresh elephant dung may hold 7,000 beetles. Within a day or two, they will have buried it all.

Some beetles roll away dung with their back legs.

▼ Turning the tables

A sundew plant feeds on insects which are trapped by sticky droplets on the stalks on its leaves. But caterpillars of a small plume moth feed on sundews in Florida, U.S.A. Detachable scales on the moth probably save it from being caught. The caterpillars drink the sticky droplets, then eat the stalks and any trapped insects.

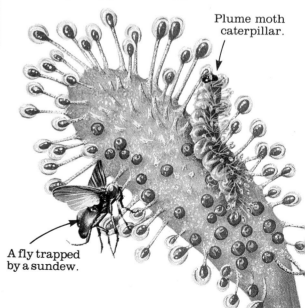

Plume moth caterpillar.

A fly trapped by a sundew.

▼ Feeding on stones

Males of some butterflies – Purple Emperor, Red Admiral and White Admiral – are sometimes seen licking dry stones on woodland tracks. They are probably taking in sodium salts. Why only males? The packet of sperms passed to the female during mating contains a rich soup of nutrients. Scientists in North America have found that some butterfly species need to make up this loss of sodium after mating.

Purple Emperor

The butterfly makes a damp patch on the stone, then sucks it dry.

Battle of the giants ▼

Large, poisonous bird-eating spiders, with legs spreading up to 20 centimetres, are preyed on by digger wasps with wing spans of 10 centimetres.

Bird-eating spider

◄ Having found a spider, the female wasp inspects it carefully, then digs a hole, keeping watch on the spider.

Digger wasp

The wasp then attacks ► the spider. As she jabs with her sting, the spider is alerted at last and fights, but too late.

◄ The sting paralyses the spider but does not kill it. The wasp drags the spider into the hole.

After laying an egg, the ► wasp buries the spider. The young wasp will eat it.

TRUE or FALSE?

In Africa there are man-eating flies.

Vampire moth ►

Of the 200,000 species of butterflies and moths, only one is known to suck blood. This innocent-looking Malaysian moth pierces the hides of tapirs, buffaloes and other mammals. It may suck blood for up to an hour. The diagrams show how its proboscis pierces the skin and then drills into the animal's flesh.

Head

Muscles

Skin

Proboscis

◄ The proboscis is in two halves. Rapid side-to-side bending drives one tip then the other against the skin, finally breaking through.

Neck muscles

Skin

Barbs

Blood pressure erects barbs to grip the flesh.

▲ Fast rocking of the head straightens the proboscis and drives it into the flesh.

Chemical warfare

Harmful chemicals are used by insects as repellents, in defence and attack. Most advertise this by their colour pattern, often some combination of black with yellow, orange or red. These colours are a universal code meaning "don't touch!"

Frothing at the mouth ▶

Flightless Lubber Grasshoppers are large and slow. When disturbed, however, they ooze foul-smelling froth from the mouth and thorax with a hissing noise. Air is bubbled into a mixture of chemicals which include phenol and quinones. Both these chemicals are widely used as repellents by insects.

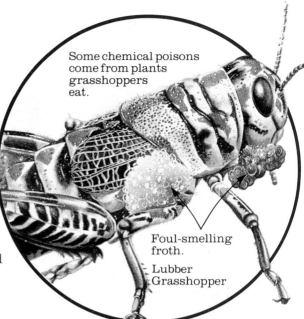

Some chemical poisons come from plants grasshoppers eat.

Foul-smelling froth.

Lubber Grasshopper

◀ Painful jabs

The stings of ants, bees and wasps are modified ovipositors (egg-laying tools), used to inject poison in defence or to paralyse prey. More than 50 different chemicals have been identified from various species. Some cause itching, pain, swelling and redness; others destroy cells and spread the poison. Honeybees cannot pull their barbed stings from human skins, and tear themselves away, dying soon afterwards.

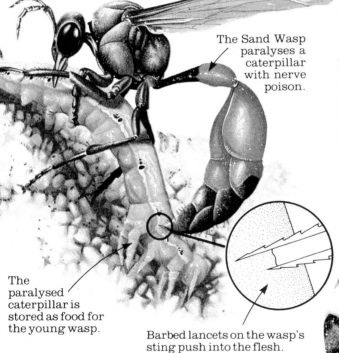

The Sand Wasp paralyses a caterpillar with nerve poison.

The paralysed caterpillar is stored as food for the young wasp.

Barbed lancets on the wasp's sting push into the flesh.

Bombardment ▶

When provoked, a bombardier beetle swivels the tip of its abdomen and shoots a jet of boiling chemicals at its attacker. The chemicals are produced in a "reaction chamber" with an explosion you can hear. The spray of foul-tasting, burning vapour is a result of rapid firing. It shoots out at 500 to 1,000 pulses per second at a temperature of 100°C.

Bottoms up ▶

Darkling beetles respond to trouble by doing a hand-stand. They tilt up at an angle of 45° and point their abdomen at the attacker. They then spray a foul-smelling liquid from glands that open at the tip of the abdomen. Darkling beetles are slower than bombardier beetles and are often swallowed by toads before discharging their spray.

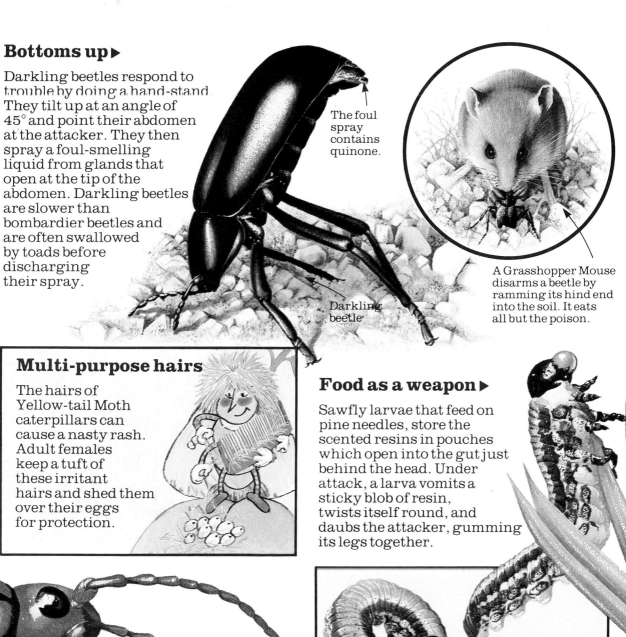

The foul spray contains quinone.

Darkling beetle

A Grasshopper Mouse disarms a beetle by ramming its hind end into the soil. It eats all but the poison.

Multi-purpose hairs

The hairs of Yellow-tail Moth caterpillars can cause a nasty rash. Adult females keep a tuft of these irritant hairs and shed them over their eggs for protection.

Food as a weapon ▶

Sawfly larvae that feed on pine needles, store the scented resins in pouches which open into the gut just behind the head. Under attack, a larva vomits a sticky blob of resin, twists itself round, and daubs the attacker, gumming its legs together.

A sawfly larva gumming up an ant.

The bombardier beetle's spray can be fired accurately in any direction.

Ladybirds bite when annoyed.

TRUE or FALSE?

Curious courtship

Insects have many ways of attracting and recognising a mate, of arousing their interest, calming their fears, and overcoming their aggression.

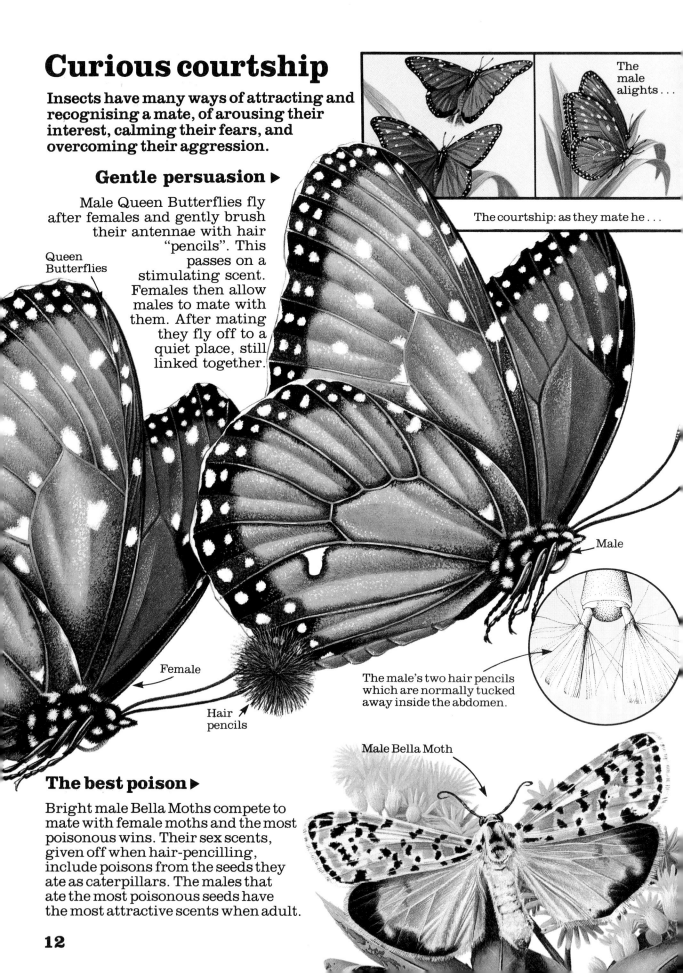

The courtship: as they mate he . . .

The male alights . . .

Gentle persuasion ▶

Male Queen Butterflies fly after females and gently brush their antennae with hair "pencils". This passes on a stimulating scent. Females then allow males to mate with them. After mating they fly off to a quiet place, still linked together.

Queen Butterflies

Male

Female

Hair pencils

The male's two hair pencils which are normally tucked away inside the abdomen.

Male Bella Moth

The best poison ▶

Bright male Bella Moths compete to mate with female moths and the most poisonous wins. Their sex scents, given off when hair-pencilling, include poisons from the seeds they ate as caterpillars. The males that ate the most poisonous seeds have the most attractive scents when adult.

...they mate...

...and fly off together.

...strokes her antennae as if to calm her.

TRUE or FALSE?

Singing crickets attract more females.

Off with his head!

In some species of preying mantis, the female begins to eat the male while they are mating. She starts at his head and by the time she reaches his abdomen, mating is completed. By becoming a nourishing meal, the father provides a supply of food for the eggs that are his children.

Bribing the lady ▼

Courtship and mating are dangerous for males if females are insect-eaters. Many male Empid flies distract their females with the gift of a captured insect. This stops the female eating the male. Some wrap the "gift" in silk.

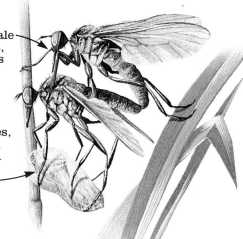

While the female Empid fly eats, the male mates with her.

In nectar-feeding species, males offer an empty balloon of silk — a ritual gift.

Fatal attraction ▼

Male and female fireflies recognise and find each other by light signals. Each species has its own pattern of flashes. Predatory female *Photuris* mimic the signals of female *Photinus* to lure male *Photinus* to their deaths. As their male prey approaches, they take off and "home in" like light-seeking missiles.

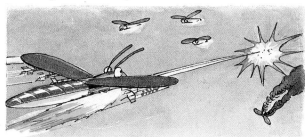

Female firefly

A female flashing in answer to a male.

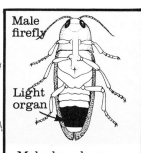

Male firefly

Light organ

Males have larger lights than females.

Magical changes

Grasshoppers and several other sorts of insects hatch out of eggs as miniature adults, except for their wings and reproductive organs which develop later.▶

In many other groups of insects, such as beetles and flies, the young insect (the larva) is quite unlike the adult. The larva concentrates on feeding and growing before turning into a pupa. Great changes of the body take place in the pupa to produce the winged adult.▼

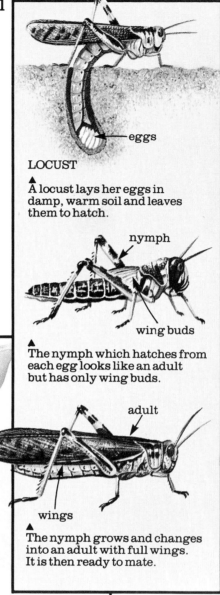

LOCUST
▲
A locust lays her eggs in damp, warm soil and leaves them to hatch.

nymph

wing buds
▲
The nymph which hatches from each egg looks like an adult but has only wing buds.

adult

wings
▲
The nymph grows and changes into an adult with full wings. It is then ready to mate.

eggs

▼ Egg machines

Moth caterpillars, known as bag-worms, build a protective case of silk and fragments of twigs and leaves. They pupate in the case, and the wingless females never leave. Many females are worm-like, with no legs, eyes or mouthparts. Their only function is to mate and lay eggs. Males are normal. Three of the many different bag-worms are shown here.

A North American bag-worm caterpillar eating leaves.

A case becomes a cocoon.

A male Malayan bag-worm mating. The female is still inside her case.

LADYBIRD

eggs →

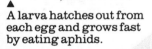

▲
After mating, the ladybird (a beetle) lays her eggs on a leaf.

larva

▲
A larva hatches out from each egg and grows fast by eating aphids.

The larva changes ▶ into a pupa. An adult ladybird forms inside the pupa's skin.

pupa

adult

◀When the pupa's skin splits an adult ladybird emerges, ready to find a mate.

A mobile nest ▶

Some giant water bugs lay their eggs on the male's back, sticking them on with waterproof "glue". Female giant water bugs often leave the water and fly about but males stay until their eggs have hatched out. This "giant" is only 2-5 cm long but many are larger: they lay eggs on water plants.

The male "nursemaid"

Beauty and the beast ▶

The nymphs of dragonflies are fearsome underwater predators, even eating small fish. By pumping water in and out of its rectum, where its gills are, a nymph can move by jet propulsion.

A nymph eating a tadpole, caught by hooks on its lower lip.

A fully-grown nymph climbs out of the water and a dragonfly emerges.

Adult dragonfly

Coming out together

The periodical cicadas of the eastern United States spend 17 years (13 in the south) below ground as nymphs feeding on tree roots. All in one place emerge together. They change into adults, lay eggs, and after a few weeks die. None is seen again for 17 (or 13) years.

TRUE or FALSE?

Butterflies live for only one day.

A red aphid giving birth. Other daughters cluster behind her.

Nymphs hatch from the eggs.

Inside a Russian doll is a smaller doll and within that is a still smaller doll, and so on . . .

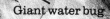

Giant water bug

Russian dolls ▲

For most of the year, female aphids produce young without mating. Eggs develop into small daughters inside the mother. Inside each daughter eggs also start to develop. So when a mother gives birth to her daughters, they already contain her grand-daughters, like a set of dolls.

Royal households

The social life of two different groups of insects: ants, bees and wasps, and termites, has evolved separately. Most nests have only one egg-laying female, known as the queen. The teeming members of a nest are her offspring. Most are workers who never breed. The queen keeps her ruling position and controls the nests' activities by chemical communication.

The queen honeybee is at the centre of the swarm.

To fill her crop with nectar, a bee worker visits up to 1,000 flowers. She flies 10 such trips a day if it is sunny.

To make 1 kg of honey, bees make up to 65,000 trips, visiting 45-64 million flowers.

Large colonies have 80,000 workers and eat 225 kg of honey a year. Surplus honey is stored for bad weather.

Honeycomb cells are six-sided.

Deciding where to go ▶

A new honeybee nest starts when thousands of workers, all female, leave the old nest with the old queen, or with some virgin queens, when a few males (drones) go too. Before they fly off, scout bees find suitable nesting sites and report back, telling of their finds by "dancing". After several hours, by unknown means, the site for the new nest is chosen.

Bees swarm on the ground, on branches and even on post-boxes in towns.

TRUE or FALSE?

Bees make jellies.

Nest bullies ▲

Dominant *Polistes* worker biting a subordinate.

Bullying ends in food exchange.

Dominant *Polistes* wasp workers bully their sister workers. A dominant one will bite a subordinate, as she crouches motionless, until she regurgitates her food. This seems to bind the colony together.

Incredible industry ▶

Vespula wasps build their football-sized nests with paper made of chewed-up wood mixed with saliva. A large nest has 12 combs containing 15,000 cells, surrounded by walls of layered paper. Unlike the honeybees, wasps abandon their nests in autumn and build a new one in spring, sometimes underground in an old animal burrow.

Wasps do not make honey but feed their larvae on pellets made of chewed-up insects.

◀ Equipped for the job

As a worker honeybee grows older, her body changes for different tasks. Her salivary glands start to produce "brood food" for the larvae on the fifth or sixth day. This stops by the twelfth day and glands on her abdomen start to produce wax. Such tasks largely depend on a bee's age although the time-table is not rigid. Workers live for 6-8 weeks. Queens can live for up to 5 years.

DAY 1

DAY 6

DAY 12

DAY 19

DAY 26

Workers start by cleaning the hive, then feed the young and go on to make and repair wax cells, guard the hive and, finally, to visit flowers for pollen and nectar.

Honeybee's wax glands.

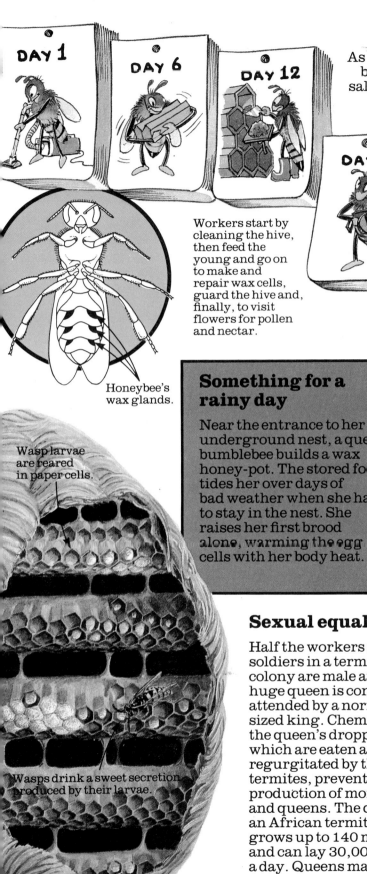

Wasp larvae are reared in paper cells.

Wasps drink a sweet secretion produced by their larvae.

Something for a rainy day

Near the entrance to her underground nest, a queen bumblebee builds a wax honey-pot. The stored food tides her over days of bad weather when she has to stay in the nest. She raises her first brood alone, warming the egg cells with her body heat.

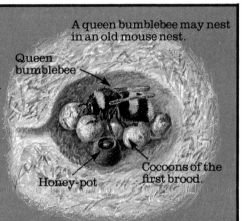

A queen bumblebee may nest in an old mouse nest.

Queen bumblebee

Honey-pot

Cocoons of the first brood.

Sexual equality ▶

Half the workers and soldiers in a termite colony are male and the huge queen is constantly attended by a normal-sized king. Chemicals in the queen's droppings, which are eaten and regurgitated by the termites, prevent the production of more kings and queens. The queen of an African termite species grows up to 140 mm long and can lay 30,000 eggs a day. Queens may live for 15 years and more.

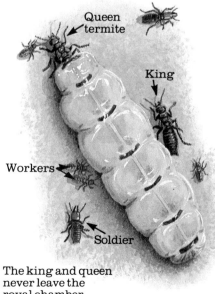

Queen termite

King

Workers

Soldier

The king and queen never leave the royal chamber.

Farmers, tailors, soldiers and builders

Large insect colonies, particularly those of termites and ants, have different members which are specialized for different types of jobs. By working together efficiently, they have developed unusual ways of using and feeding on plants and other animals.

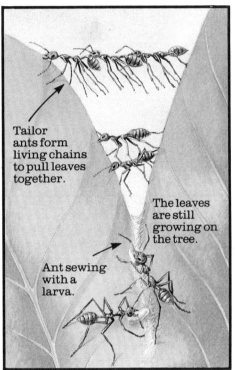

Tailor ants form living chains to pull leaves together.

The leaves are still growing on the tree.

Ant sewing with a larva.

◀ Tailor ants

Oecophylla ants sew leaves together to form a nest, using silk produced by the saliva glands of their larvae. A line of workers stand on the edge of one leaf, and pull another towards it with their jaws. Other workers wave larvae to and fro across the leaves until they are joined by the silk. Tailor ants live in the forests of Africa, S.E. Asia and Australia.

The abdomen may be 1 cm in diameter.

Honey-pot ant

A section cut through a termite mound, showing the queen's chamber, air-conditioning channels and fungus gardens.

Channels control climate and air flow.

Royal chamber

Fungus gardens

Minders ▶

When cutting or carrying bits of leaf, leaf-cutter ants cannot defend themselves against parasitic flies. Tiny workers, too small to cut and carry leaves, go with the larger workers to fight off the flies. The ants carry the leaf pieces back to their nest.

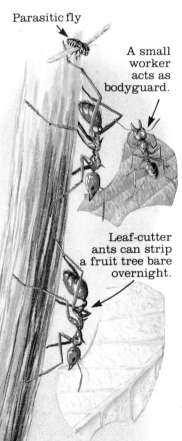

Parasitic fly

A small worker acts as bodyguard.

Leaf-cutter ants can strip a fruit tree bare overnight.

Once in the nest the leaves are chewed up, mixed with saliva and droppings, and used as a compost for the fungus which the ants feed on.

The fungus gardens break down plant material. Termites feed on the white blobs.

A worker feeding.

Honey-pot ants are a sweet delicacy to Mexican villagers and Australian aborigines. An ant colony may have up to 300 living "honey-pots".

Living storage tanks ▲

Some honey-pot ant workers are fed so much nectar and honeydew that their abdomens swell to the size of a grape. Unable to move, they hang motionless from the roof of the nest and are looked after by other workers. The "honey-pots" store food for the colony to eat when the nectar season is over in the deserts where they live. When empty they shrivel up.

Soldiers and weapons

Soldier termites have enlarged heads and defend the nest against intruders. Most have formidable pointed jaws as weapons, but not *Nasutitermes* soldiers. Their heads extend forwards as a pointed nozzle from which they spray a sticky, irritating fluid.

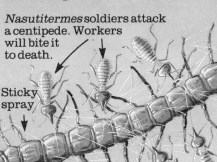

Nasutitermes soldiers attack a centipede. Workers will bite it to death.

Sticky spray

Master builders

Fungus garden termites build enormous, rock-hard mounds with sand, clay and saliva. One vast mound contained 11,750 tonnes of sand, piled up grain by grain. The nest may be in or below the mound and the shape varies with species, soil type and rainfall. Some mounds are 9 m tall

Some mounds are like pagodas, with roofs to shed rain.

Mounds of the Australian Compass Termite all face the same way. The broad sides, facing east-west, catch maximum heat from the weak sun at dawn and dusk.

▼ Laying in a store

Harvester termites cut grass into short pieces and store it in their warm, damp underground nests where temperatures are constant. Harvester ants bring seeds from the desert floor into their underground granaries. Husked and chewed into "ant bread", it provides food during shortages.

Fresh grass is stored in chambers near the surface. When dry it is then moved close to the nest.

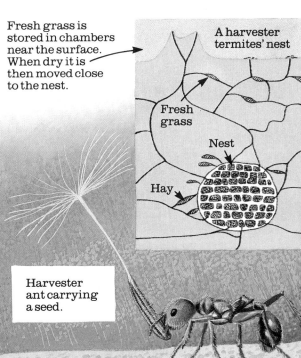

A harvester termites' nest

Fresh grass

Nest

Hay

Harvester ant carrying a seed.

Recognition and deception

Scent, sight and sound are used by insects to identify each other and to pass on information to members of the same species. They may also be used for defence and disguise.

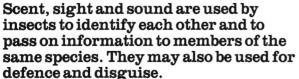
A colony of aphids.

Lacewing larva covered with aphid wax.

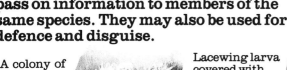

Tiger moth

Flexing this cuticle makes a clicking noise.

The ear picks up a bat's clicks.

Bright colours deter daytime predators.

Noises in the night

Many nasty-tasting tiger moths make high-pitched clicks at night. Bats, which hunt by echo-location, veer away from clicking moths. The Banded Woolly Bear Moth is quite edible but it clicks like other tiger moths and so is left alone by the bats.

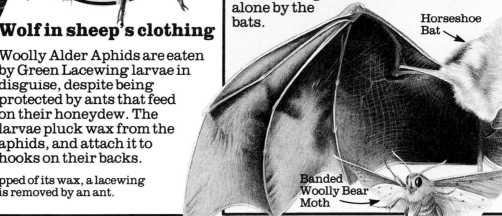

Horseshoe Bat

Banded Woolly Bear Moth

Wolf in sheep's clothing

Woolly Alder Aphids are eaten by Green Lacewing larvae in disguise, despite being protected by ants that feed on their honeydew. The larvae pluck wax from the aphids, and attach it to hooks on their backs.

If stripped of its wax, a lacewing larva is removed by an ant.

Turning up the volume

Mole crickets broadcast their songs from specially built burrows. The Y-shaped burrow acts as an amplifier so that the call can be heard from further away.

Noses on stalks ▶

An insect's antennae carry dozens of tiny structures which are sensitive to scent. They are also sensitive to touch. Antennae are used both to smell and to touch and stroke. They provide an insect with much information.

Cockchafers spread their antennae to find food.

Antennae

Emperor Moth

Grasshopper band ▶

Male grasshoppers stridulate, or sing, by rubbing a "file" across a "scraper". They mainly sing to attract female grasshoppers.

A line of pegs on the inside of the "thigh" acts as the file. They rub across a hard vein on the wing.

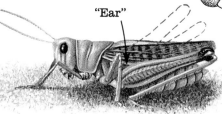

"Ear"

Close-up of pegs.

Short-horned grasshoppers rub the hind legs against the wings. Their "ears" are on the side of the body.

Scraper

File

A scraper on one wing rubs against a file on the other.

"Ears" are in their front "knees".

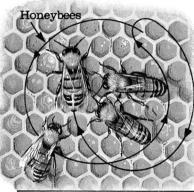

Long-horned grasshoppers (like the Spike-headed Katydid on the cover) rub their wings together.

TRUE or FALSE?

Crickets have thermometers.

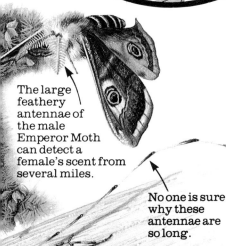

Dancing bees ▼

Honeybees tell others where to find food by "dancing" on the comb. Sound, scent and food-sharing also pass on information about the type of food and where it is.

Round dance
A round dance means food within 80 m. The richness of the source is shown by the energy and length of dance. Other bees pick up the scent of the flowers.

Honeybees

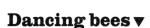

Waggle dance

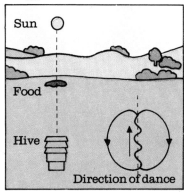

Sun

Food

Hive

Direction of dance

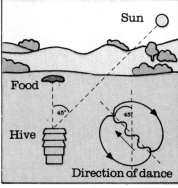

Sun

Food

45°

Hive

45°

Direction of dance

The large feathery antennae of the male Emperor Moth can detect a female's scent from several miles.

No one is sure why these antennae are so long.

Timberman Beetle

A waggle dance describes food over 80 m from the hive. Speed and number of waggles on the straight run indicate distance. The angle between

straight run and vertical shows the direction of the food relative to the sun. As the sun moves across the sky, the angle of dance changes.

Living – and dying – together

Many insects live with and make use of a wide variety of other animals and plants. Some even share ants' nests and many are "milked" of sweet liquids by ants. These relationships have led to special ways of feeding, finding shelter and breeding. Sometimes the association is good for both sides. More often, one makes use of the other, usually by eating it.

Special delivery ▼

Human Botflies use biting flies, such as mosquitoes, to deliver their eggs to human, bird or mammal hosts. Grubs hatch from the eggs and burrow into the host's skin.

A female botfly catches a mosquito and glues a cluster of 15-20 eggs on the mosquito's body.

When the mosquito settles on a person, the eggs hatch immediately in response to warmth.

Over 100 moths, of seven species, have been found in the coat of one sloth.

Three-toed Sloth

Moths and sloths ▲

Sloths are so sluggish that moths infest their long coats which are green with algae. About once a week, a sloth slowly descends its tree to excrete. Its moths lay eggs in the dung, which the caterpillars eat. Since their "home" never moves far or fast, moths have no trouble finding it again.

Velvet ant

Solitary wasp pupa.

◄ Pretty but painful

Velvet ants are not ants. They are really densely hairy, wingless, female wasps. Most break into the nests of solitary bees or wasps to lay their eggs on a pupa, which the velvet ant grub will eat. The brilliant colours of the wasps perhaps warn that they have a sting so powerful they are called "cow-killers".

22

A houseful of lodgers ▶

Robin's pincushions on wild rose are galls caused by a tiny wasp which lays eggs in young leaf buds. After the gall has formed, another gall wasp enters and lays its eggs. The larvae of both owner and lodger are parasitized by chalcid and ichneumon wasps, and these in turn may also be parasitized.

The original Robin's pincushion gall wasp.

Gall wasp larvae inside the pincushion.

TRUE or FALSE?

Springtail "jockeys" ride soldier termites.

Treacherous guest ▼

Although they eat ant larvae, caterpillars of the Large Blue Butterfly are allowed to live in the nests of *Myrmica* ants. They produce a sweet fluid which the ants love.

Tiny caterpillars eat Thyme flowers. Older ones will produce a sweet fluid for an ant, which responds by carrying it into its nest.

Caterpillar in ant nest.

Ants' nest.

Adult Large Blue

Parasites on parasites

The Mexican Bean Beetle eats leaves and is a pest in North America. It is parasitized by a tachinid fly, which is itself parasitized by an ichneumon wasp.

Mexican Bean Beetle

Tachinid fly

Ichneumon wasp

Beetle's larva

Fly larva inside beetle larva

Wasp laying eggs in the fly's pupa

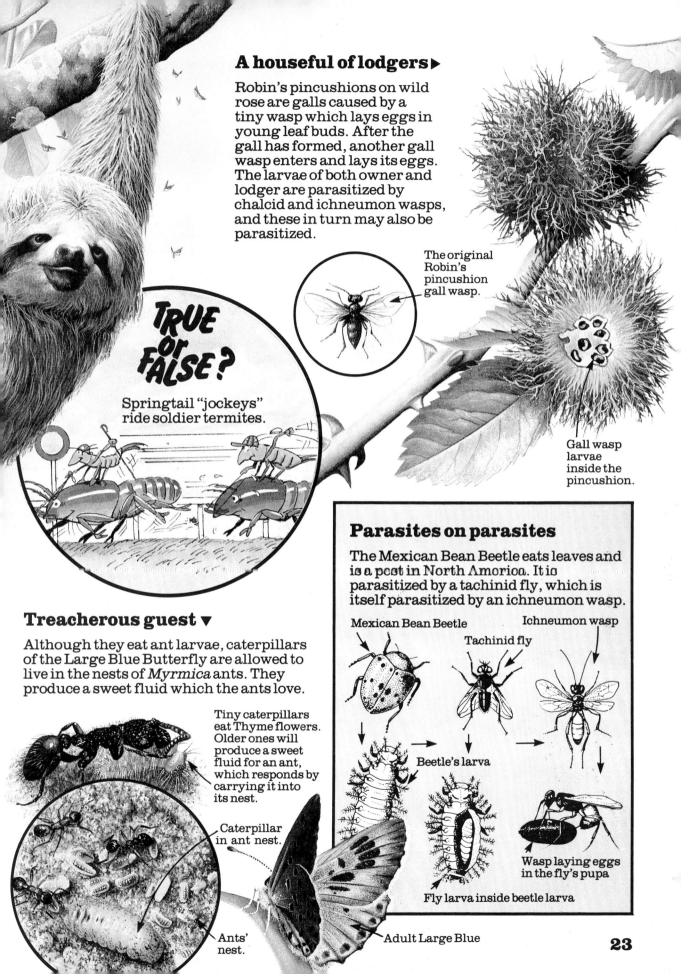

Travellers and hitchhikers

Insects not only make long journeys under their own power but are also carried, often accidentally, by people or other animals.

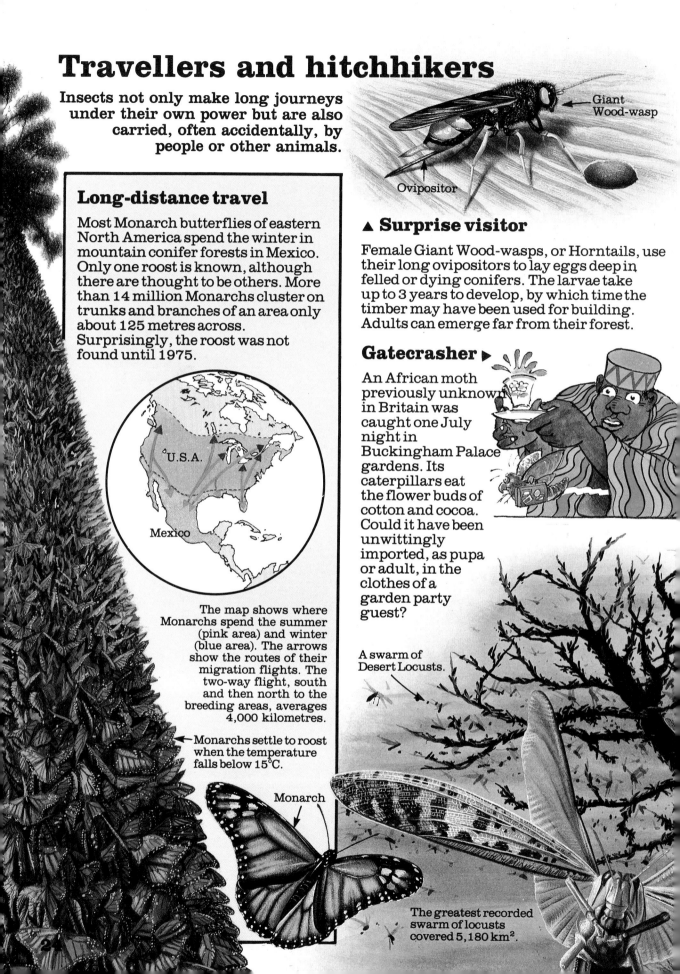

Giant Wood-wasp

Ovipositor

Long-distance travel

Most Monarch butterflies of eastern North America spend the winter in mountain conifer forests in Mexico. Only one roost is known, although there are thought to be others. More than 14 million Monarchs cluster on trunks and branches of an area only about 125 metres across. Surprisingly, the roost was not found until 1975.

U.S.A.

Mexico

The map shows where Monarchs spend the summer (pink area) and winter (blue area). The arrows show the routes of their migration flights. The two-way flight, south and then north to the breeding areas, averages 4,000 kilometres.

← Monarchs settle to roost when the temperature falls below 15°C.

Monarch

▲ Surprise visitor

Female Giant Wood-wasps, or Horntails, use their long ovipositors to lay eggs deep in felled or dying conifers. The larvae take up to 3 years to develop, by which time the timber may have been used for building. Adults can emerge far from their forest.

Gatecrasher ▶

An African moth previously unknown in Britain was caught one July night in Buckingham Palace gardens. Its caterpillars eat the flower buds of cotton and cocoa. Could it have been unwittingly imported, as pupa or adult, in the clothes of a garden party guest?

A swarm of Desert Locusts.

The greatest recorded swarm of locusts covered 5,180 km².

Drifts of butterflies ▶

White butterflies in the drier parts of Africa fly off in huge clouds at the start of the dry season when there is no food for the caterpillars. One observer in East Africa reported 500 million butterflies passing each day on a 24 kilometre wide front. Migrating butterflies are sometimes blown out to sea and then washed up on beaches.

TRUE or FALSE? Young beetles hitch lifts on bees.

▼ Locust plagues

From time to time Desert Locusts breed rapidly and millions fly long distances in search of food. An area of 26 million km², from West Africa to Assam, and Turkey to Tanzania is at risk of invasion. A large swarm eats 80,000 tonnes of grain and vegetation a day.

Far from water

Large numbers of *Sympetrum* dragonflies fly south in autumn across France and through mountain passes in the Pyrenees. The exact route of their migration is not known but some turn up in Portugal. On northward migrations the dragonflies often reach south-eastern England from Europe.

Migrating dragonflies

Mulberry Silkworm Moth

The caterpillar is called a silkworm.

Caterpillars are fed on mulberry leaves, and pupate in a cocoon spun from a single thread of silk over 900 m long.

Cocoon

Home-grown moths ▲

Silkworm Moths have been reared in China for their silk for over 4,000 years. The moth's wings are so small it cannot fly and it no longer exists in the wild.

The Chinese kept silk-making a secret. One story says that it did not leave China until 350 AD when a princess hid eggs in her head-dress when she went to India to marry a prince.

Citizens of the world

Wherever you go on land you find insects. They can live in the harshest climates, eat whatever there is, and adapt quickly to change.

TRUE or FALSE?

Insects make long-distance phone calls.

Drinking fog ▶

Darkling beetles manage to live in the hottest, most barren parts of the Namib desert. When night-time fogs roll in from the sea, *Onymacris* beetles climb to the top of sand dunes and, head down, stand facing into the wind. Moisture condenses on ridges on their backs and runs down into their mouths.

A flightless crane fly

Mount Kilimanjaro

Staying on top of the world ▶

Above 5,000 metres on Kilimanjaro, just below the snow cap, is a bleak, windy, alpine desert, with little vegetation. Moths, beetles, earwigs, crane flies and grasshoppers live there but many are wingless. Flying insects might be in danger of blowing away.

Hot as your bath ▶

In the USA's Yellowstone National Park, famous for its geysers, nymphs of the Green-jacket Skimmer dragonfly live in hot spring water at 40°C.

Hissing Cockroaches push air through holes, hissing as they fight.

Cockroaches nibble at book-bindings, photographic film, starched linen, leather goods and any food, fouling them with strong-smelling droppings.

Easy to please

Tropical cockroaches have spread worldwide as scavengers in people's heated houses. They can and do eat anything of animal or plant origin, and this accounts for their success.

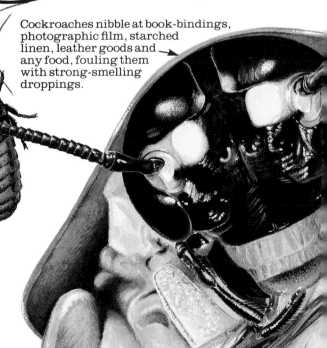

This cave cockroach from Trinidad feeds on bat droppings.

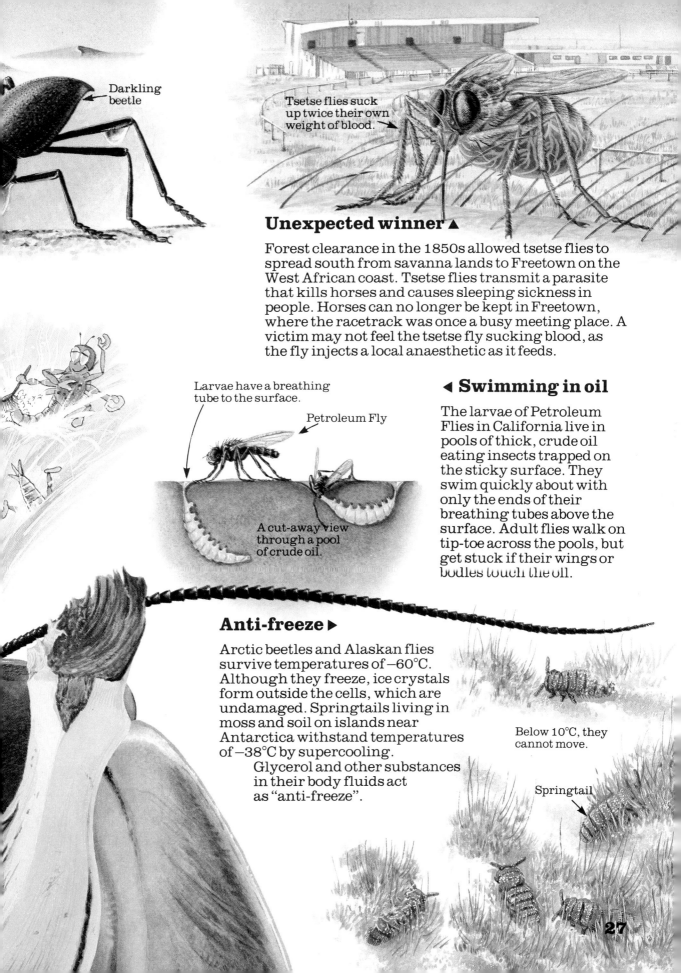

Darkling beetle

Tsetse flies suck up twice their own weight of blood.

Unexpected winner ▲

Forest clearance in the 1850s allowed tsetse flies to spread south from savanna lands to Freetown on the West African coast. Tsetse flies transmit a parasite that kills horses and causes sleeping sickness in people. Horses can no longer be kept in Freetown, where the racetrack was once a busy meeting place. A victim may not feel the tsetse fly sucking blood, as the fly injects a local anaesthetic as it feeds.

Larvae have a breathing tube to the surface.

Petroleum Fly

A cut-away view through a pool of crude oil.

◄ Swimming in oil

The larvae of Petroleum Flies in California live in pools of thick, crude oil eating insects trapped on the sticky surface. They swim quickly about with only the ends of their breathing tubes above the surface. Adult flies walk on tip-toe across the pools, but get stuck if their wings or bodies touch the oil.

Anti-freeze ▶

Arctic beetles and Alaskan flies survive temperatures of −60°C. Although they freeze, ice crystals form outside the cells, which are undamaged. Springtails living in moss and soil on islands near Antarctica withstand temperatures of −38°C by supercooling. Glycerol and other substances in their body fluids act as "anti-freeze".

Below 10°C, they cannot move.

Springtail

Insect mysteries

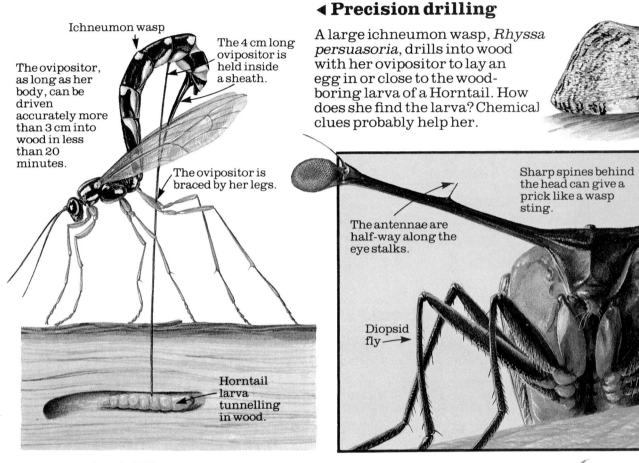

Ichneumon wasp

The ovipositor, as long as her body, can be driven accurately more than 3 cm into wood in less than 20 minutes.

The 4 cm long ovipositor is held inside a sheath.

The ovipositor is braced by her legs.

Horntail larva tunnelling in wood.

◀ Precision drilling

A large ichneumon wasp, *Rhyssa persuasoria*, drills into wood with her ovipositor to lay an egg in or close to the wood-boring larva of a Horntail. How does she find the larva? Chemical clues probably help her.

Sharp spines behind the head can give a prick like a wasp sting.

The antennae are half-way along the eye stalks.

Diopsid fly

Samson's riddle ▼

In the story of Samson, in The Bible, Samson found a swarm of bees and honey in the body of a lion he had earlier killed. Could the bees have been drone flies – a type of hoverfly which mimics honeybees? Drone flies, unlike bees, sometimes breed in rotting carcasses.

Drone flies

"Out of the eater came something to eat and out of the strong came something sweet."

Unlike most weevils, this one has a short snout.

This weevil's scientific name is *Tribus attelabini*.

Lantern Bugs are harmless suckers of tree sap.

◀ Miniature monsters

The huge heads of Lantern Bugs were once thought to be luminous which is how they got their name. Why do they look so odd? Could a monkey mistake this 10 cm long bug for an alligator?

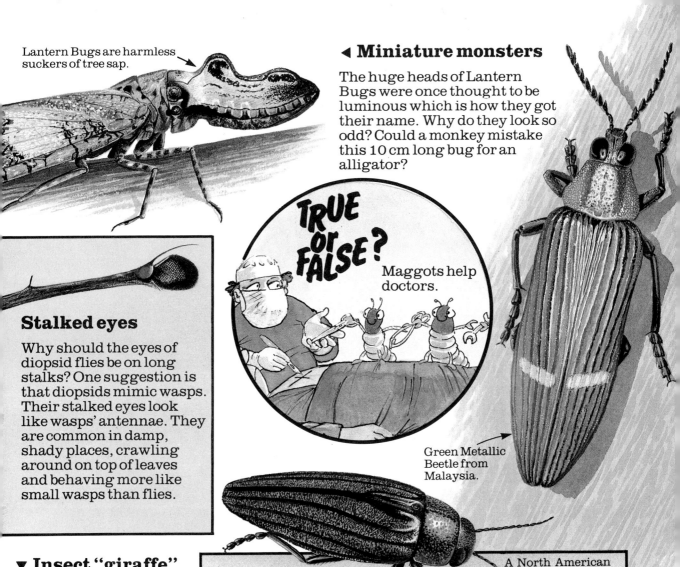

TRUE or FALSE?

Maggots help doctors.

Green Metallic Beetle from Malaysia.

Stalked eyes

Why should the eyes of diopsid flies be on long stalks? One suggestion is that diopsids mimic wasps. Their stalked eyes look like wasps' antennae. They are common in damp, shady places, crawling around on top of leaves and behaving more like small wasps than flies.

A North American metallic wood-boring beetle.

Precious beetles

Amongst the most brilliantly coloured insects are metallic wood-boring beetles but the purpose of their colour is unclear. Their shining wing-cases have been used in jewellery and embroidery. Some people have even worn tethered beetles as brooches. Toasted larvae of the Green Metallic Beetle are a delicacy.

▼ Insect "giraffe"

This odd weevil from the island of Madagascar, off Africa, has a very long head, rather than a long, giraffe-like neck. No one is quite sure why it is this shape. When threatened, it drops down and plays dead. (It is closely related to the red weevil on the front cover.)

It is 2.5 cm long and the head takes up almost half this length.

Warning mimicry? ▶

On hearing a foot-step, worker termites foraging beneath dead leaves, vibrate their abdomens. When there are many of them, the rustling of leaves sounds like the hiss of a snake.

SSSSSS

Record breakers

Loudest insect

Male cicadas produce the loudest insect sound, by vibrating ribbed plates in a pair of amplifying cavities at the base of the abdomen: this can be heard over 400 m away.

Largest and smallest butterflies

Queen Alexandra's Birdwing from New Guinea has a wingspan of 28 cm; that of a blue butterfly from Africa is only 1.4 cm.

Population explosion

If all the offspring of a pair of fruit flies survived and bred, the 25th generation — one year later — would form a ball of flies that would reach nearly from the earth to the sun.

Killer bees

So-called killer bees are an African strain of honeybee, which tend to attack people and other animals. More than 70 deaths occurred in Venezuela recently in 3 years, the number of stings varying from 200 to over 2,000. In 1964, a Rhodesian survived 2,243 stings.

Fastest flight

The record has been claimed for a hawk moth flying at 53.6 km/h, but it probably had a following wind. A more reliable record is 28.57 km/h for a dragonfly *Anax parthenope*.

Fastest runner

Try to catch a cockroach and it seems very elusive. They are among the fastest runners, reaching 30 cm per second, but this is only 1.8 km/h.

Biggest group

The insect order Coleoptera (beetles) includes nearly 330,000 species, about a third of known insects. Roughly 40,000 are weevils.

Shortest life

An aphid may develop in 6 days and live another 4-5 days as an adult. Mayflies have the shortest adult life, many living only one day after emerging from the water.

Oldest group

The oldest fossils of winged insects, dating from more than 300 million years ago, include cockroach wings.

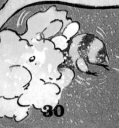

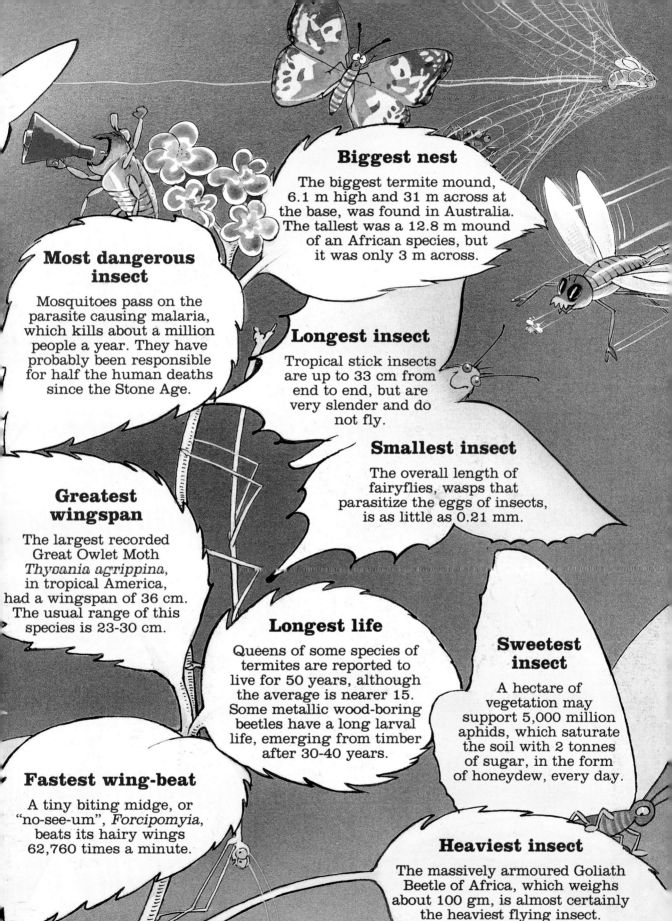

Biggest nest

The biggest termite mound, 6.1 m high and 31 m across at the base, was found in Australia. The tallest was a 12.8 m mound of an African species, but it was only 3 m across.

Most dangerous insect

Mosquitoes pass on the parasite causing malaria, which kills about a million people a year. They have probably been responsible for half the human deaths since the Stone Age.

Longest insect

Tropical stick insects are up to 33 cm from end to end, but are very slender and do not fly.

Smallest insect

The overall length of fairyflies, wasps that parasitize the eggs of insects, is as little as 0.21 mm.

Greatest wingspan

The largest recorded Great Owlet Moth *Thysania agrippina*, in tropical America, had a wingspan of 36 cm. The usual range of this species is 23-30 cm.

Longest life

Queens of some species of termites are reported to live for 50 years, although the average is nearer 15. Some metallic wood-boring beetles have a long larval life, emerging from timber after 30-40 years.

Sweetest insect

A hectare of vegetation may support 5,000 million aphids, which saturate the soil with 2 tonnes of sugar, in the form of honeydew, every day.

Fastest wing-beat

A tiny biting midge, or "no-see-um", *Forcipomyia*, beats its hairy wings 62,760 times a minute.

Heaviest insect

The massively armoured Goliath Beetle of Africa, which weighs about 100 gm, is almost certainly the heaviest flying insect.

Were they true or false?

page 5 Bumblebees have central heating.
TRUE. Bumblebees can maintain a temperature of 30-37°C when the air is near freezing. Heat is produced by a chemical process in the flight muscles.

page 7 Zebras have black and white fleas.
FALSE. A Zebra's fleas are not camouflaged.

page 9 In Africa there are man-eating flies.
TRUE. Maggots of the "tumbu" fly burrow into human skin causing painful open sores.

page 11 Ladybirds bite when annoyed.
FALSE. Blood oozes from their knee joints when they are molested, and this irritates sensitive skins.

page 13 Singing crickets attract more females.
TRUE. Males of a North American cricket attract females by "singing" in groups. Some males never sing but join the group and try to mate.

page 15 Butterflies live for only one day.
FALSE. Some live many months, hibernating in winter, or migrating long distances.

page 16 Bees make jellies.
PARTLY TRUE. The "brood food" produced by worker honeybees is called "royal jelly" because it is given only to future queens.

page 21 Crickets have thermometers.
PARTLY TRUE. The warmer it is, the faster they function. Add 40 to the number of chirps a Snowy Tree Cricket gives in 15 seconds, and you get the temperature in degrees Fahrenheit.

page 23 Springtail "jockeys" ride soldier termites.
TRUE. A West African springtail rides on a soldier's head and snatches its food.

page 25 Young beetles hitch lifts on bees.
TRUE. The larvae of some oil and blister beetles swarm over flowers and attach themselves to bees. Carried to a solitary bee's nest, they invade a brood cell and eat the egg and stored food.

page 26 Insects make long-distance 'phone calls.
FALSE. However, termites frequently bore into underground telephone cables; moisture seeps in, insulation breaks down, and the cable no longer carries messages.

page 29 Maggots help doctors.
PARTLY TRUE. As a result of experience gained in the 1914-18 war, surgeons used maggots of Green-bottle Flies, reared in sterile conditions, to clean infected wounds.

Index